The Scientific Method

A Bulletin Board in a Book!

Sunflower

e d u c a t i o n

Exceptional Books for Teachers and Parents

Editorial
Sunflower Education
Rubio Edits

Design
Cynthia Hannon Design

ISBN-13: 978-1-937166-17-5
ISBN-10: 1-937166-17-1

Table of Contents

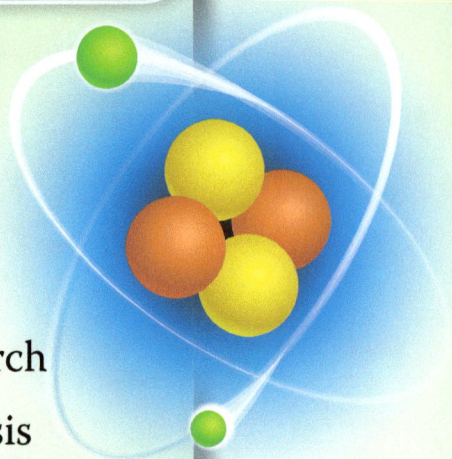

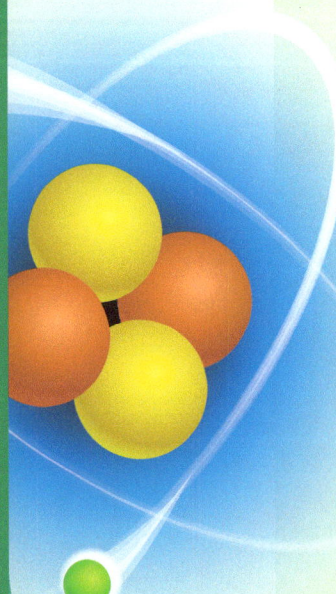

To the Teacher

The Scientific Method: A Bulletin Board in a Book! consists of two main parts: bulletin-board posters and student activity sheets. They are designed to be used together.

There are 11 posters:

- 1—The Scientific Method (title poster **A**)
- 1—Science is Knowledge poster (**B**)
- 1—7 Steps of the Scientific Method poster (**C**)
- 7—Steps of the Scientific Method posters
- 1—Remember the Scientific Method! poster (**D**)

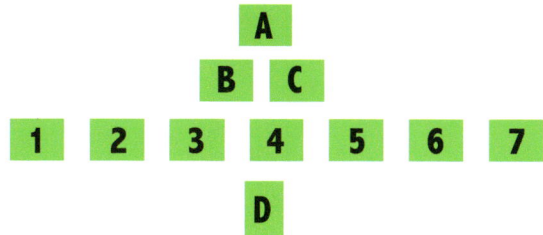

There are 8 student activity sheets:

- 1—on the scientific method in general
- 7—step-specific

❶ Post the Bulletin-Board Display

- Copy the posters or cut them out.
- See the illustration for suggested layout.
- Post the Steps posters in numerical order.

❷ Discuss the Scientific Method with Students

- Lead a discussion about the scientific method. Allow students time to peruse the posters. Ensure students understand the concepts of "if/then" statements, null hypothesis, independent and dependent variables, and controls. Share level-appropriate information about each of the steps in the scientific method:

1. Ask a Question
You can ask any question, from, "Why are there no more dinosaurs?" to, "Can human beings fly if they have wings?" Isaac Newton developed his theory of gravity after asking the question, "Why did this apple fall to the ground?"

2. Do Background Research
Some good places to do background research include books, websites, and magazines and journals. You can also talk to teachers and other appropriate adults.

3. Construct a Hypothesis
The hypothesis is based on what you learn while doing background research. It is often stated as an "if/then" statement." A hypothesis is only testable (that is, valid) if its opposite can be shown to be true: "If I drop a ball, then it will not fall to the ground." The opposite of the hypothesis is called the null hypothesis.

4. Perform an Experiment
Usually, an experiment includes an independent variable (something the scientist changes) and a dependent variable (something that will change as a result of changing the independent variable). Ideally, only one independent variable will be included in an experiment. Many experiments include a control, in which the scientist changes nothing.

5. Gather Data
This information should help you answer your Question (Step 1) by supporting or refuting your Hypothesis (Step 3). Data must be measurable. Mass, distance, and volume are measurable kinds of data. Happiness, understanding, and love are difficult to measure.

6. Analyze the Results and Draw a Conclusion
Appropriate questions: Is the hypothesis true or partly true? Is the hypothesis false? Is the null hypothesis true? What does it mean if the hypothesis is only partly true? Finding out your hypothesis was wrong is important and useful. Don't be discouraged if you learn you were wrong. Science is about knowledge, not being correct all the time!

7. Publish the Results
Publishing is important because it makes your data part of the great pool of scientific knowledge. Other scientists can use your data as part of their background research.

- Encourage students to memorize the order of the steps of the scientific method. Work with students to complete the Remember the Scientific Method! poster. Explain that the letters are the first letter of the main idea of each step and that they are listed in order of occurrence. Encourage students to come up with a memorable phrase (for example, "Queen Rachel Hopes Every Dog Receives Praise") as a mnemonic device.
- Focus students' attention on the relation between steps (such as Step 2: Do Background Research, and Step 7: Publish the Results). Encourage students to think critically about each step and the reasons for it.

❸ Share the Activity sheets

- All of the activity sheets can be completed using information from the posters and critical thinking skills. Consider having students complete the activity sheets either as assessment or with access to the bulletin board.
- Students can complete the activity sheets individually or with partners.

Emphasize that the scientific method is a process used to learn about the world and that students are scientists when they follow it.

Have fun exploring science with your students!

Worksheet Answers

The Scientific Method

1. The steps scientists follow to advance knowledge.
2.

Step	Description
1. Ask a Question	Ask something you want to learn about
2. Do Background Research	Gather information about what other scientists have already learned
3. Construct a Hypothesis	Write an "if/then" statement to answer the question
4. Perform an Experiment	Conduct a step-by-step process that tests the hypothesis
5. Gather Data	Collect measurable information to help answer your question by supporting or refuting your hypothesis
6. Analyze the Results and Draw a Conclusion	Determine if your hypothesis was true
7. Publish the Results	Share the information for other scientists to use when they do the research step

Step 1: Ask a Question

1. Ask something you want to learn about.
2. This step is the reason for the process.
3. Reward insightful answers.

Step 2: Do Background Research

1. Gather information about what other scientists have already learned.
2. Determine if the answer to your question has already been determined, or, if not, what it is likely to be.
3. Others must publish their results in order for you to have information to research.

Step 3: Construct a Hypothesis

1. Write an "if/then" statement to answer the question.
2. Without this step, you cannot know if your question was answered.
3. This step provides the testable answer to the question.

Step 4: Perform an Experiment

1. An experiment is a step-by-step process that tests the hypothesis.
2. Verify correct answers.

Step 5: Gather Data

1. Data is measurable information that helps you answer your question.
2. Collect measurable information to help answer your question by supporting or refuting your hypothesis.
3. Verify correct answers.

Step 6: Analyze the Results and Draw a Conclusion

1. Determine if your hypothesis was true.
2. This step organizes data into patterns that support or refute the hypothesis to determine the answer to the question.
3. Results are the data and patterns that they show, while the conclusion is the interpretation of what those results mean.

Step 7: Publish the Results

1. Share the information for other scientists to use when they do the research step.
2. Without this step, knowledge is not advanced.
3. Reward insightful answers.

The Scientific Method

1 What is the scientific method?

_____

_____

2 List the steps of the scientific method. Describe each one.

Step	Description
1.	
2.	
3.	
4.	
5.	
6.	
7.	

Step 1: Ask a Question

1 Describe this step.

_____

_____

_____

_____

2 Why is this step important?

_____

_____

_____

_____

3 How do you come up with questions to ask?

_____

_____

_____

_____

Step 2: Do Background Research

1 Describe this step.

_____

_____

_____

_____

2 Why is this step important?

_____

_____

_____

_____

3 How is this step related to Step 7: Publish the Results?

_____

_____

_____

_____

Step 3: Construct a Hypothesis

1 Describe this step.

_____

_____

_____

_____

2 Why is this step important?

_____

_____

_____

_____

3 How does this step relate to Step 1: Ask a Question?

_____

_____

_____

_____

Step 4: Perform an Experiment

1 What is an experiment?

_____

_____

_____

_____

2 Tell about an experiment you have done or heard about.

_____

_____

_____

_____

_____

_____

_____

_____

_____

Step 5: Gather Data

1 What is data?

_____

_____

_____

_____

2 Describe this step.

_____

_____

_____

_____

3 Give an example of data you have gathered in order to answer a question.

_____

_____

_____

_____

Step 6: Analyze the Results and Draw a Conclusion

1 Describe this step.

_____

_____

_____

_____

2 Why is this step important?

_____

_____

_____

_____

3 What is the difference between results and a conclusion?

_____

_____

_____

Step 7: Publish the Results

1 Describe this step.

_____

_____

_____

_____

2 Why is this step important?

_____

_____

_____

_____

3 Where are some places you might publish your results?

_____

_____

_____

_____

The Scientific Method

Science is Knowledge

- The word *science* comes from from the Latin word *scientia*, which means "knowledge."
- Science gives us knowledge about the world.
- Science has been the basis of the progress of humanity for centuries.

To advance knowledge, scientists follow the 7 steps of the scientific method:

1. Ask a Question
2. Do Background Research
3. Construct a Hypothesis
4. Perform an Experiment
5. Gather Data
6. Analyze the Results and Draw a Conclusion
7. Publish the Results

① Ask a Question

What do you want to learn about?

② Do Background Research

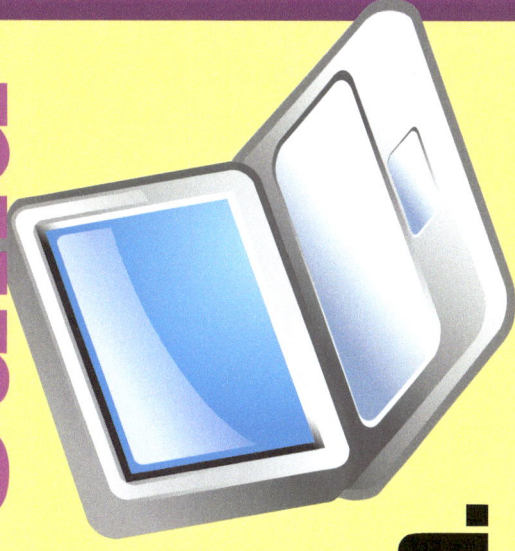

Gather information. What have other scientists already learned?

③ Construct a Hypothesis

An educated guess of the answer to the question.

④ Perform an Experiment

A step-by-step process that tests the hypothesis.

5 Gather Data

Data is information collected during the experiment.

6 Analyze the Results
Draw a Conclusion

- The results are the pattern or trend shown by the data.

- The conclusion is your interpretation of what those results mean.

⑦ Publish the Results

Share the information for other scientists to use when they do the research step.

Remember the Scientific Method!

Q _____

R _____

H _____

E _____

D _____

R _____

P _____